ISO 14001
Step by Step

A practical guide

Second edition

ISO 14001
Step by Step

A practical guide

Second edition

NAEEM SADIQ

ASIF HAYAT KHAN

IT Governance Publishing

Every possible effort has been made to ensure that the information contained in this book is accurate at the time of going to press, and the publishers and the author cannot accept responsibility for any errors or omissions, however caused. No responsibility for loss or damage occasioned to any person acting, or refraining from action, as a result of the material in this publication can be accepted by the publisher or the author.

Apart from any fair dealing for the purposes of research or private study, or criticism or review, as permitted under the Copyright, Designs and Patents Act 1988, this publication may only be reproduced, stored or transmitted, in any form, or by any means, with the prior permission in writing of the publisher or, in the case of reprographic reproduction, in accordance with the terms of licences issued by the Copyright Licensing Agency. Enquiries concerning reproduction outside those terms should be sent to the publishers at the following address:

IT Governance Publishing Limited
Unit 3, Clive Court
Bartholomew's Walk
Cambridgeshire Business Park
Ely
Cambridgeshire
CB7 4EA
United Kingdom
www.itgovernancepublishing.co.uk

The authors have asserted the rights of the author under the Copyright, Designs and Patents Act, 1988, to be identified as the authors of this work.

First published in the United Kingdom in 2011 by IT Governance Publishing.
ISBN 978-1-84928-102-7

Second edition published in the United Kingdom in 2019 by IT Governance Publishing.
ISBN 978-1-78778-032-3

FOREWORD

Individuals, organisations and communities across the globe have become increasingly concerned with the quality and sustainability of our environment. There is also a growing demand for organisations to demonstrate environmentally responsible behaviour through eco-friendly products, processes and practices.

This book is for all those who wish to develop and implement an effective ISO 14001-based environment management system (EMS). International standards, such as ISO 14001, often carry a formality about them with each word bearing a specific meaning and context. They define requirements on 'what' ought to be done, without prescribing any specific details on 'how' they might be achieved. There is, therefore, a need to demystify ISO 14001 by presenting its requirements, as well as an implementation methodology, in a simple, user-friendly and easily understandable manner.

This practical guide is ideal for managers, auditors and trainers who are involved in any aspect of an ISO 14001-based management system – whether development, implementation, training or auditing. We will take a hands-on and step-by-step approach, explaining the purpose and the requirement of each clause, and how an organisation can meet those requirements. This guide also includes numerous examples, suggestions and documentation samples to facilitate understanding and implementation for those who are not familiar with this subject.

ABOUT THE AUTHORS

Naeem Sadiq holds a BSc in aerospace and a master's in manufacturing engineering. He is a certified lead auditor and lead trainer for the ISO 9001, ISO 14001 and OHSAS 18001 standards. He is also an ASQ-certified manager and quality system auditor.

Naeem's work experience in engineering and management includes working as an independent consultant, auditor and trainer for the ISO 9001, ISO 14001 and OHSAS 18001 standards.

He has presented a number of papers in national conferences on management system standards, and provided consultancy, training and auditing support to more than 100 organisations. As a freelance writer, he is a regular contributor to a national newspaper on environmental and social issues. He is also the author of *OHSAS 18001 Step by Step – A Practical Guide*.

Asif Hayat Khan holds a BSc in industrial engineering. He is a trained lead auditor and trainer for the ISO 9001, ISO 14001 and OHSAS 18001 standards. He is also an ASQ-certified quality manager, HACCP auditor and a DuPont certified trainer on behavioural safety.

As a consultant, auditor, trainer and manager, he has experience of establishing and implementing the ISO 9001, ISO 14001 and OHSAS 18001 standards in diverse industrial sectors.

ACKNOWLEDGEMENTS

We deeply acknowledge all those organisations and people with which we have had the privilege of working, training, consulting and auditing. These rich and varied experiences helped our own learning and understanding, without which this book would not have been possible.

DISCLAIMER

All names, examples and values quoted in this book are fictitious and have been presented for learning, understanding and explaining purposes only.

Websites quoted in this book may change over time and users may need to look up new addresses where a change has taken place.

CONTENTS

CHAPTER 1: PDCA, ENVIRONMENTAL POLICY, OBJECTIVES AND PROGRAMMES (CLAUSES 5.1, 5.2 AND 6.2)

First principles: the Plan-Do-Check-Act (PDCA) model

The PDCA model provides the iterative process organisations use to ensure continual improvement takes place. As ISO 14001 puts it, PDCA forms the "basis for the approach underlying an environmental management system".

The idea is that you first establish exactly what it is you intend to do (Plan) – what are the environmental objectives and processes necessary to stay in line with environmental policy? After that, you implement planned processes (Do). Next, you monitor and measure these processes against the environmental policy (Check) and report the results. Finally, you take corrective action based on these results (Act), thus continually improving.

Before beginning anything else, the organisation needs to ensure that it understands this process because it will apply to every part of the EMS. The environmental policy, for instance, will need to be subject to reviews and improvements in the same way that other processes that could have environmental impact are.

Environmental policy

An environmental policy is a statement of an organisation's top management commitment, defining the direction and intentions of the organisation with regard to its environmental performance. The policy must be consistent with the nature, scale and environmental impacts of the organisation's activities, products and services – and, naturally, approved by top management.

Summary of requirements

The following diagram provides an overview of what an EMS policy statement must contain:

Figure 1: EMS policy requirements

How can this requirement be met?

The following points define what an organisation must do to meet the requirements of the Standard:

- Top management must document and approve an environmental policy that reflects its vision and commitment to being an environmentally responsible organisation.
- The EMS policy must be communicated to all employees. An organisation can choose many methods to communicate its policy, such as training sessions, video messages, displays at prominent locations, newsletters or embedded in emails.
- Top management must review the EMS policy at planned intervals to ensure its continued suitability and adequacy.

An example of an environmental policy is shown below:

Eco-Friendly Inc (EFI) environmental policy

EFI is committed to providing fast-moving home-use products in an environmentally responsible manner. The organisation will fulfil its environmental commitment by:

- Assessing the environmental risks and operating the business in a way that ensures prevention of pollution through the application of economically viable, best-available environmental practices;
- Ensuring compliance with applicable environmental legislation;

- Collaborating with suppliers for sustainable sourcing of raw materials, as well as with the transporters, carriers, business partners and other concerned organisations for improving end-to-end environmental performance; and
- Continually improving products, processes and ways of doing business that reduce levels of environmental impact through sustainable initiatives, such as energy, water and natural resources conservation, waste and gaseous emissions reduction (particularly the greenhouse gases), and exploring opportunities for reuse and recycling.

EFI identifies and reviews environmental objectives in annual workshops with management and the board.

Objective, targets and programmes

Summary of requirements

- The organisation is required to set environmental performance goals and targets.
- The organisation shall establish and implement programmes for achieving its objectives and targets.

Considerations when setting environmental objectives and targets

While setting objectives and targets, the organisation must consider the following:

- **EMS policy:** top management's commitment and vision, as stated in the organisation's EMS policy.
- **Significant environmental aspects:** these are the environmental aspects of the organisation's activities, and may include factors over which the organisation has limited control (e.g. secondary impacts as a result of a supplier's activities, or where a given process is the only known way to perform an activity).
- **Legal and other requirements:** the organisation will likely be subject to a range of legal, regulatory and contractual obligations relating to its environmental impact and activities. These may need to be addressed through objectives.
- **Technological options and financial considerations:** naturally, the organisation should consider technological options and financial viability as it establishes its EMS objectives.
- **Operational and business requirements and views of interested parties:** these are further factors to consider when establishing EMS objectives. Environmental concerns, such as coastal and marine life, banning chromium from the leather industry, using environmentally friendly dyes in textiles and garments, minimising the use of non-biodegradable plastic, and minimising noise, smoke or traffic congestion, often raised by interested parties ought to be considered while establishing EMS objectives. They may also include considerations regarding the

local community or environment, or be closely related to the organisation's defined culture and business objectives.

Some example EMS objectives and targets are shown in *Table 1*.

Table 1: EMS Objectives and Targets

Objectives	Targets
Removal of ozone-depleting substances from operations	100% removal by 2025
Reduction in water consumed per tonne of production	Reduce from current consumption level of 5 m^3/tonne to 4 m^3/tonne of production by December 2021
Effluent treatment plant expansion to cater for enhanced production volume projected for 2023	Existing capacity = 10 tonnes per hour Target = 15 tonnes per hour by December 2023

Environmental programmes

An environmental programme is a roadmap or a plan for achieving an environmental objective. It defines responsibilities, means and timeframes by which the objectives are to be achieved. An example of an environmental programme is given in *Table 2*.

Table 2: Example Environmental Programme

Objective: reduce water consumption from the current baseline of 15,000 gallons per month per 100 tonnes of production, to 12,000 gallons per month per 100 tonnes of production by July 2021			
No.	**Actions/projects**	**Responsibility**	**Target date**
1.	Replace hosepipe floor cleaning with vacuum cleaner	Admin Manager	March 2020
2.	Recycle treated and grey water for factory plantation	Engineering Manager	Dec. 2020
3.	Install auto-taps in the washrooms	Engineering Manager	Jan. 2021
4.	Replace shower system for cleaning of product with dip cleaning	Engineering Manager	June 2021

Measurement, monitoring and changes to environmental objectives, targets and programmes

The organisation is required to monitor the extent to which its environmental objectives and targets have been met. This information must be fed to top management for review of the EMS. This information may be needed to determine whether:

- Objectives and targets are being achieved as planned;
- Additional resources are necessary; and
- The planned objectives, targets and programmes have to be adjusted.

CHAPTER 2: IDENTIFYING ENVIRONMENTAL ASPECTS AND DETERMINING SIGNIFICANT IMPACTS (CLAUSE 6.1.2)

Summary of requirements

Establish and implement procedure(s) to:

- Identify environmental aspects of all activities, products and services that are performed within the defined scope of the organisation;
- Determine those that have or can have significant impacts on the environment; and
- Establish an EMS, which must be based on consideration of the significant environmental aspects.

Terms used

Environment:

surroundings in which an organisation operates, including air, water, land, natural resources, flora, fauna, humans and their interrelationships[1]

Because of its products, processes and activities, an organisation constantly interacts with its surroundings.

---

[1] ISO 14001:2015, Clause 3.2.1.

The complete set of surroundings in which an organisation operates may be considered its environment.

Environmental aspect:

element of an organisation's activities or products or services that interacts or can interact with the environment[2]

An organisation may interact with environments in many ways. These could relate to the activities it carries out, the products it makes and the processes it follows to make these products. Any such component that interacts with the environment is considered an environmental aspect of the organisation.

There are two types of environmental aspects:

1. **Direct environmental aspects:** those aspects over which an organisation can be expected to influence and control, such as emissions from its processes.
2. **Indirect environmental aspects:** those aspects over which the organisation can be expected to influence, but not control, such as energy consumption for the production of raw materials or emissions from trucks used by suppliers.

How can these requirements be met?

- Begin by establishing a cross-functional environmental aspect and impact assessment team.

---

[2] ISO 14001:2015, Clause 3.2.2.

Team members must be familiar with environmental issues, laws, the processes under study and the procedure for determining aspects and impacts.

- Document a procedure that defines the complete process for identification of aspects and impacts. A sample procedure is shown in *Appendix A*.
- Make an inventory of all processes, products and activities (routine, non-routine, normal, abnormal and emergency) performed by the organisation.
- Identify inputs and outputs of each process, product and activity.
- Identify environmental aspects of each input and output using a black-box approach, as shown in *Figure 2*.

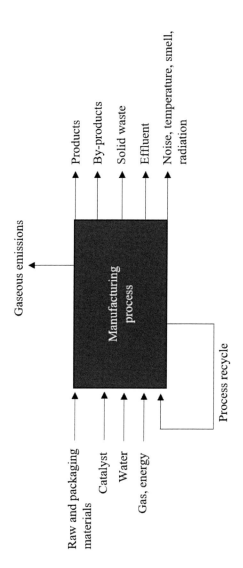

Figure 2: Identifying environmental aspects – the input-output approach

- The environmental aspects could be:
  - Emissions to air
  - Release to water;
  - Waste management and disposal;
  - Contamination of land;
  - Impact on communities;
  - The use of raw materials, energy and natural resources;
  - The use of ozone-depleting or radioactive materials; and/or
  - Other local environmental and community issues.
- Determine the environmental impact of each aspect. When doing so, consider controls already in place, severity, frequency, legal requirements and reputational impact. A sample environmental significance rating system is described in *Appendix A*.
- Document aspects and impacts in a register of significant environmental aspects (see *Table 4* of *Appendix A* for a sample format).
- Review the register of significant environmental aspects as an ongoing activity (say, annually). Also review the register when new aspects are identified, when an environmental incident takes place, or whenever processes and activities are added or modified.
- Consider significant environmental aspects while establishing the organisation's EMS. The significant

environmental aspects should form the basis of the EMS objective-setting process. The significant environmental aspects must also be considered while establishing other elements of the environmental system, such as training, operational controls, measuring and monitoring, structure and responsibility, emergency preparedness, audits and management reviews.

- Consider environmental aspects and their significance while designing new products and processes. Such assessments should also be conducted before making any changes to existing processes and activities.

CHAPTER 3: REGULATORY REQUIREMENTS AND EVALUATING COMPLIANCE (CLAUSES 6.1.3 AND 9.1.2)

Summary of requirements

Establish and implement a system that:

- Defines the process of identifying and accessing applicable requirements (relating to its environmental aspects);
- Determines how those requirements apply;
- Considers these requirements while establishing and implementing the EMS; and
- Periodically verifies the organisation's continued compliance with these requirements.

How can these requirements be met?

The first step in managing compliance with environmental legal requirements is to know which requirements apply to the environmental aspects of an organisation's activities, products and services.

Each country has its own environmental laws and regulatory bodies. The three environmental regulators in the UK are: the Environment Agency for England and Wales (EA), the Scottish Environment Protection Agency (SEPA) and the Environment and Heritage Service Northern Ireland (EHS NI). A good source of regulatory information is NetRegs (*www.netregs.gov.uk*), a web-based tool developed by these three organisations.

Laws relating to local authorities, such as land use planning, local air quality strategies (to reflect the national air quality strategy), Local Authority Air Pollution Control (LAAPC), clean air acts, noise and statutory nuisance and tree preservation orders should also be reviewed for possible applicability.

Regulators often implement environmental regulations by issuing environmental permits that contain conditions for the permit holder to fulfil. These are normally applied to operations at specific sites or installations, and typically regulate emissions to air (e.g. local authority air pollution control), water (e.g. consents to discharge) and land (e.g. waste management licences). Pollution Prevention and Control (PPC) permits are used to regulate all media on an integrated basis.

It is the responsibility of an organisation to apply for an environmental permit that is applicable and required under the law.

Clause 6.1.2 of ISO 14001 requires an organisation to define and implement a procedure that describes how it identifies and implements the following activities:

- Identify all environmental laws, conventions, permits and licences that apply to its products, processes, outputs, liquid discharges, solid waste, emissions and activities. Not every regulatory requirement applies to all organisations. An organisation needs to determine the requirements that apply to its own products, processes and activities. A list of such laws and requirements should be made and kept updated.

- Identify how an organisation accesses the sources (departments, ministries, websites, groups, organisations) that create or issue this information. The purpose of this requirement is to ensure that an organisation can access all sources on an ongoing basis to keep itself abreast of the latest regulatory and other applicable requirements.

- Regulatory requirements play an important role in defining an EMS. An organisation needs to consider its legal requirements while establishing its environmental objectives and programmes. Training, measuring, monitoring, communication, organisational structure, operational controls and responsibilities are the other elements that will normally be influenced by applicable environmental laws.

- ISO 14001 requires that an organisation must periodically evaluate (periodicity may be defined by law or by the organisation) its compliance against all applicable environmental laws and other requirements. The mechanism and responsibility for performing this task must be defined by the organisation.

Websites that could provide useful regulatory information about applicable environmental laws in the UK, Germany,

US, Canada, Australia and China are listed in the table below:

Table 3: Websites that provide useful information about applicable environmental laws

Country	Organisation	Website
UK	NetRegs	*www.netregs.gov.uk*
	The Environment Agency	*www.gov.uk/government/organisations/environment-agency*
	Environmental Protection UK	*www.environmental-protection.org.uk*
	Department for Environment Food & Rural Affairs (DEFRA)	*www.gov.uk/government/organisations/department-for-environment-food-rural-affairs*
Germany	Umwelt Bundesamt	*www.umweltbundesamt.de*

US	Environmental Protection Agency	*www.epa.gov*
Canada	Environment and Climate Change Canada	*www.canada.ca/en/environ ment-climate-change.html*
Australia	Department of the Environment and Energy	*www.environment.gov.au*
China	Ministry of Ecology and Environment	*www.mee.gov.cn*

Records to be kept

The following records must be maintained:

- An updated list of all applicable requirements.
- The results of measuring, monitoring and testing, carried out at a defined frequency to demonstrate that the organisation complies with all applicable requirements.

CHAPTER 4: RESOURCES, ROLES, RESPONSIBILITY, AUTHORITY AND COMMUNICATION (CLAUSES 5.3, 7.1 AND 7.4)

Summary of requirements

- Provide sufficient resources to establish, implement, maintain and improve the EMS.
- Define and assign roles and responsibilities, and appoint specific management representative(s) for the EMS.
- Establish procedure(s) that define processes for internal and external communication.

How can these requirements be met?

An EMS system cannot be implemented or improved unless management provides the necessary human, financial, material, specialised skills and organisational resources. This may be done in the following manner:

1. Senior management defines the organisation's structure, and the roles, responsibilities and authorities of all those who are required to perform any task relating to EMS. These could be tasks relating to identifying significant environmental aspects; establishing EMS objectives; ensuring the application of EMS controls; planning, measuring and monitoring EMS performance; training;

communicating with internal and external bodies; ensuring compliance; and so on.

2. The next step is to ensure that the individuals have the skills and resources to perform their assigned roles. (See *Chapter 5* for competency requirements.)

3. Top management must nominate a competent person as a 'management representative'. They must have the authority to ensure that the EMS is established, implemented and maintained according to the requirements of ISO 14001. They must also report the performance of the EMS to top management for review and further improvement. (See *Chapter 12* on elements that constitute this report.) Some aspects of a management representative's competence could include knowledge of environmental issues, ISO 14001, environmental controls and good communication skills.

Communication

To implement an effective EMS, an organisation must communicate internally as well as externally.

Internal communication

This focuses on communication within the organisation and could entail:

- Awareness of and training on all EMS issues, requirements and procedures at various levels within the organisation;

- Communicating applicable requirements to all concerned;
- Receiving feedback on the measuring and monitoring of the EMS's performance;
- Communicating the results of the EMS's performance to top management;
- Ensuring that people at all levels and in all functions receive the needed EMS information in a reasonable timeframe; and
- Ensuring that employees are able to forward suggestions and EMS concerns to the relevant management.

The following list provides examples of internal communication:

- Bulletin board displays about the key EMS goals and the extent to which they have been achieved.
- Regular EMS meetings, at various levels of management and for various levels of employees to discuss and review EMS issues.
- Training and awareness sessions.
- Reporting EMS performance to top management.
- Emails and notes in salary slips informing employees about EMS initiatives.
- Suggestion boxes.
- Newsletters.

External communication

An organisation may need to communicate with external stakeholders. These could be customers, vendors, suppliers, contractors, neighbours, environmental groups and regulators.

External communication could entail:

- Receiving and responding to concerns of interested parties, such as NGOs, environmental groups and communities;
- Communicating with regulatory and environmental protection agencies; and/or
- Communications relating to permits, licences and fines.

A sample communication procedure can be found in *Appendix B*.

The organisation must choose whether it wishes to communicate externally about its significant environmental aspects. If it decides to do so, it must define who will communicate, to whom they will communicate and how this communication will take place.

Examples of external communication:

- Maintaining a website that describes the organisation's EMS policy and initiatives.
- Meetings with vendors and suppliers to explore options for environmentally friendly products.

- Informing customers about the organisation's EMS initiatives and encouraging them to do the same.
- Participating in community meetings.
- Receiving and responding to government and regulatory agencies.

CHAPTER 5: COMPETENCE, TRAINING AND AWARENESS (CLAUSES 7.2 AND 7.3)

Summary of requirements

- People performing tasks that relate to potentially significant environmental impacts must have the knowledge and skills to perform those tasks.
- Identify and provide training associated with the EMS and the organisation's environmental aspects.
- Establish procedure(s) to provide awareness relating to significant aspects, the EMS policy, objectives and procedures, including the potential consequences of not complying with these procedures.

How can these requirements be met?

Clauses 7.2 and 7.3 describe how an organisation can ensure the competence of all personnel who perform tasks for or on behalf of the organisation with the potential to have a significant environmental impact. The following step-by-step approach, backed up by a formal procedure, could effectively meet these requirements:

1. Identify all individuals whose tasks have the potential to have a significant environmental impact. *Figure 3* explains how inputs from many elements of an EMS contribute to the training needs analysis process.

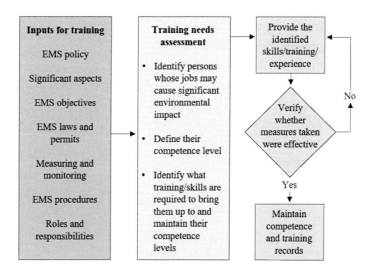

Figure 3: Training needs assessment

2. The main thrust of these clauses is to ensure that all personnel are aware of their environmental roles and responsibilities, have the knowledge and skills to perform them effectively, and understand the potential consequences of not following procedures. This should include both employees and third parties such as contractors.

3. Make a list of all EMS-related tasks throughout the organisation with the potential to have a significant environmental impact. The environmental tasks considered should include those performed routinely and in emergency situations.

4. It is important to verify whether the training provided, or other actions taken, actually achieved the intended goal(s). There are many methods that may be used to determine the effectiveness of actions taken. These could range from a simple examination for theoretical subjects to practical testing for skill-based subjects.

5. Keep records of training provided. These could relate to training needs analyses, training calendars, attendance sheets, certificates or records of training effectiveness.

CHAPTER 6: DOCUMENTATION AND DOCUMENT CONTROL (CLAUSE 7.5)

There has traditionally been a distinction between what we call a 'document' and what we call a 'record'. Although ISO 14001 no longer really recognises this distinction (both are "documented information"), they are still useful terms to appreciate.

Documents provide information, data or instructions and could be on any medium, such as print or electronic. Some examples of documents are:

- Policies, objectives, targets, programmes or plans;
- Manuals and procedures;
- Work instructions or standard operating procedures;
- Photographs, videos or drawings providing instructions;
- Legal requirements; and
- Formats and checklists (when blank).

Records are a subcategory of documents. They provide evidence, results or facts relating to actions that have been taken by an organisation. Some examples of records are:

- Training records;
- Driver's licence;
- Machine maintenance records;
- Results of effluent, emission or solid waste discharge tests;

- Audit results or nonconformity reports;
- Minutes of management review meetings; and
- Photographs, videos or drawings providing evidence (flight data recorder, CCTV images, and so on).

ISO 14001 now treats both 'documents' and 'records' as **'documented information'**, with identical requirements.

Summary of requirements

- The organisation shall establish the documented information required by ISO 14001, as well as that relating to its own significant environmental aspects.
- The mandatory documented information is listed in *Appendix D*.

How can these requirements be met?

EMS scope

Organisations need to define the scope of their EMS in order to establish what is included in the management system. Typically, the EMS's scope will include:

- All the organisation's activities that take place on its premises;
- Areas that are under the organisation's direct control; and
- Areas where the organisation has environmental regulatory liability.

The scope may also include areas that are outside the organisation's direct control but over which it may be able to exert some influence, such as encouraging employees to use carpooling or mass transportation systems when travelling to and from work.

An example of an organisation's EMS scope could be:

> The EMS applies to the Eco-Friendly Inc facilities at 300 Minnesota Avenue, Kansas City, Kansas, to all remote locations within Kansas, and to activities carried out by Eco-Friendly Inc employees when acting on behalf of Eco-Friendly Inc. Eco-Friendly Inc is engaged in the manufacture of fast-moving home use products.

Risks and opportunities

In many management systems, this refers to a process of risk assessment and management. In the case of ISO 14001, this involves assessing the risks that the organisation faces, such as an incident causing a leak of industrial chemicals, terrorist attack, misuse of equipment, and so on. It also asks the organisation to consider the opportunities it has, such as having access to more markets if the organisation meets certain environmental standards, or long-term savings from moving to an initially more expensive but more reliable and more environmentally friendly process.

The organisation should assess the risks and opportunities in relation to the organisation's context (its industry, partners and other stakeholders, legal and regulatory obligations, and so on), its environmental aspects and its compliance obligations. The organisation will need to document the process it uses to develop this assessment as well as the results and any information about actions the organisation is taking in response. This is especially important as a number of decisions will be based on these results.

Evidence of activities

A documented EMS is only useful if it is followed, so the organisation will need to make sure it tracks compliance with the management system and records evidence. This evidence will also show where there are weaknesses or inefficiencies, and can be used both internally and externally:

- Externally to demonstrate compliance with the Standard and with relevant laws, regulations and contracts.
- Internally to identify weaknesses for treatment, to assess the effectiveness of remedial actions, to assess how cost-effective a given process is, and so on.

General EMS documentation

The documentation should define the main elements of the EMS and how these interact. It should also include or provide reference to related documents. Interaction of elements may be described in any suitable manner, such as describing them in words, or making flow charts or block diagrams. The core document featuring the EMS's scope, as well as the descriptions and interactions of its elements, is often referred to as the EMS manual.

EMS documents

ISO 14001 refers to two types of documented information needed for an EMS: those explicitly required by the Standard and those that the organisation determines are necessary for the EMS to be effective. The procedure for an internal EMS audit is an example of a document required by ISO 14001, while a procedure for cleaning a scrubber is an example of a document the organisation needs to effectively manage its EMS.

Control of documented information

An organisation needs documented procedure(s) for controlling its documents (Clause 7.5.3). Organisations might choose to distinguish between documents and records and define separate procedures for controlling each, or an integrated procedure covering both.

The issues addressed by a documentation control procedure are described in a sample procedure in *Appendix C.*

Appendix D describes the documents required by ISO 14001.

CHAPTER 7: OPERATIONAL CONTROLS (CLAUSE 8.1)

What are operational controls?

Operational controls may be defined as 'measures taken to manage risks'. In the context of an EMS, operational controls will be referred to as 'the methods and means implemented for managing environmental risks', where their absence could lead to:

- Deviation from the organisation's environmental policy, objectives and targets;
- Violation of specified legal or other applicable requirements; and/or
- Environmental pollution.

The following diagram explains the requirement of operational controls:

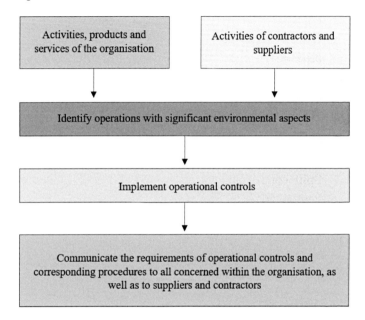

Figure 4: Operational controls

Types of operational controls

Operational controls may be broadly categorised as administrative controls or engineering controls.

Administrative controls include:

- Written procedures or instructions (operations, waste/effluent/emission/hazardous substance management, utilities, etc.);

- Supervision;
- Competence or training requirements;
- Permit to work (PTW);
- Own or contractor's environmental control procedures (induction or refresher training, meetings, inspections, access controls, PTW systems, etc.);
- Inspection and maintenance programmes;
- Solid waste disposal procedures;
- Fire watch;
- Emergency preparedness procedures and drills;
- Spill control and treatment procedures (equipment, neutralising substances, kits, etc.); and
- The use of personal protective equipment (PPE).

Engineering controls include:

- Isolation and enclosure of environmentally hazardous materials;
- Containment walls for hazardous liquids (chemicals, flammables);
- Leak detection and alarm systems;
- Electronic or mechanical interlocks to prevent emissions, overflows or spills;
- Line evacuation systems (pigging, etc.) to minimise solid waste and water consumption;
- Smoke/fire detection and suppression systems such as auto sprinklers;

- Plant safety devices (visual controls and indicators, non-return valves, pressure relief systems, etc.);
- Emergency shutdown systems;
- Effluent treatment plants; and
- Scrubbers and sprays to reduce particulate matter and other hazardous gases from emissions.

While selecting operational controls, the organisation should consider the following hierarchy:

- Elimination (e.g. lead-free petrol, chrome-free leather).
- Reduction (recycle, reuse or substitute, e.g. reducing paper thickness in packaging, recycling grey water for plantation, reusing printing toners, substituting incandescent bulbs with LED lights).
- Isolation (enclosure/segregation, e.g. building a containment wall around fuel storage, separate containers to collect different types of wastage).
- Control (through administrative and/or engineering controls).
- Defences (e.g. PPE, emergency procedures, fire-fighting, spill management).

CHAPTER 8: EMERGENCY PREPAREDNESS AND RESPONSE (CLAUSE 8.2)

Summary of requirements

The organisation shall establish and implement procedure(s) that identify potential emergency situations (that could negatively affect the environment), and define mechanisms to respond to these situations. These actions must include mitigation of any adverse environmental impacts.

The organisation must also periodically review and test these procedures.

How can these requirements be met?

There are two main components to Clause 8.2: 'preparedness' and 'response'.

Preparedness

- With a group of individuals from various functional areas in the organisation, it is important to brainstorm possible accidents and emergencies that could have environmental impacts.
- Review incident records for past years. Also review the environmental aspects list for potential emergencies under abnormal operating conditions. Environmental emergencies/accidents might include:
  o A hazardous gas leak;

- A fire or explosion;
- Unplanned emissions or effluent discharges;
- A spill;
- Failure of a tank, dam, equipment or structure;
- A natural disaster, e.g. lightning, earthquake, flood or extreme weather;
- A crash or collision;
- Sabotage, vandalism, terrorist attack, bomb threat or a riot; and
- Any other impact arising out of hazardous substances handled, stored and/or used in the organisation.

- Identify and list all potential emergencies and accidents that the organisation might encounter.
- Determine what is required to respond to these emergency situations. This could include establishing and implementing the following:
  - An emergency response plan, including roles and responsibilities.
  - The resources and equipment needed to respond to and to prevent or mitigate associated environmental impacts. These could include spill kits to contain the spread of harmful liquids, suction pumps, equipment for removal of contaminated soil, and so on.
  - Emergency procedures that must be adopted throughout the organisation.

- o Training personnel on emergency procedures and actions.
- o Emergency information and communication processes, which may include maps/drawings of electrical cabling, gas pipes, underground tanks, storm water channels, containment pits, and so on.
- o The liaison and support needed from external agencies, such as the police, fire brigade, hospitals and concerned environmental agencies.
- Locations where an emergency could occur should be mapped and have emergency equipment nearby.
- Neighbourhood areas vulnerable to gas escape or other consequences of an emergency at the facility should also be identified and marked on a map for rapid notification and other relief/mitigation actions. Prevailing wind directions should be mapped to identify the potential downwind areas. Sensitive areas in the vicinity may include residential, industrial, agricultural, recreation, fishing areas or sanctuaries.
- Emergency exits and muster points should be identified and marked.

Response

ISO 14001 requires organisations to respond to accidents or emergencies by taking actions that prevent or mitigate associated adverse environmental impacts. Effective response requires not just preparedness, but also rehearsals and drills of the procedures to make sure that:

- The procedures are adequate and can effectively respond to a real emergency or accident;
- All persons understand their roles and know how to perform them in emergency situations; and
- The emergency equipment is tested and the results of the tests are reported.

An organisation must test its emergency procedures by simulating relevant kinds of emergencies under conditions that are as realistic as possible. Such drills do not have any learning value unless they are closely monitored against their key performance parameters, e.g. response time, ability to contain a spill, fire-fighting, effectiveness of equipment, ability to evacuate and effectiveness of mitigation measures. Results of such drills must be maintained and analysed for further improvement of the emergency response system.

Organisations must review their emergency procedures at planned intervals and after an accident or emergency. This ensures the emergency preparedness and response procedures remain adequate and suitable.

CHAPTER 9: MEASURING AND MONITORING (CLAUSE 9.1)

Summary of requirements

Organisations should establish and implement procedure(s) to measure and monitor the aspects of the EMS's performance that could have significant environmental impacts.

How can these requirements be met?

The measurement and monitoring process must be viewed in a larger context and its scope extended to the entire EMS. Mechanisms and responsibilities must be defined to continuously measure and monitor key characteristics of those processes and activities with the potential for significant environmental impact.

Key characteristics provide vital information or can have significant influence on the environmental performance, which therefore need to be measured or monitored. These might include, for example:

- The amount of effluent discharged;
- The amount of biological oxygen demand or chemical oxygen demand in outgoing effluent;
- The number of spills;
- Temperature limits in a chemical warehouse;
- The amount of hazardous waste;
- Unplanned releases;

- The quantity of toxic chemicals released; or
- The amount of fuel, gas, electricity or water used, etc.

They also need to define mechanisms and responsibilities for regular monitoring/measuring of the following:

- The equipment used in emergency situations, spill kits, smoke detectors, fire detection and fire control systems.
- Controls whose absence or failure could significantly impact the environment (e.g. relief valves, one-way valves, pressure and temperature sensors, auto warning signals, pressure vessel inspections, critical inspections, permits to work, controls over suppliers, and contractors).
- The extent to which the environmental objectives and targets have been achieved.
- Chemical, fuel and hazardous waste handling, storage and disposal methods.
- Training needs, training conducted, employee awareness and competence.
- Communications from interested parties.
- Effectiveness of emergency response measures.
- Effectiveness of corrective and preventive actions.
- Environmental performance of contractors.

A measuring and monitoring plan should form the core of the measuring and monitoring process. The plan should include:

- Parameters to be measured/monitored;
- Responsibilities;
- Procedure;
- Frequency; and
- Records to be kept.

CHAPTER 10: INTERNAL AUDIT (CLAUSE 9.2)

Summary of requirements

- Establish audit procedure(s) that define the responsibilities for planning, conducting, reporting and following up internal environmental audits, and define the frequency, scope and process.
- Ensure internal EMS audits are conducted at planned intervals.
- The audits shall determine whether the EMS is in line with any planned arrangements and ISO 14001. Audits must also determine whether the EMS has been implemented effectively. The auditors shall be objective and impartial.

How can these requirements be met?

Some of the terms used in the audit process are:

- **Audits**
  Intended to determine the extent to which an organisational system complies with its own criteria. Audits must be carried out by someone not directly responsible for the activity under audit. The process of evidence collection and evaluation against the criteria must be carried out in a structured, documented and objective manner.

- **Audit criteria**

  The set of requirements against which the audit is carried out.

- **Audit findings**

  What an auditor determines after comparing and evaluating the evidence against the audit criteria. An audit finding could result in a 'conforming' or 'nonconforming' situation.

Implementing an internal EMS audit process requires the following:

- Establish an internal EMS audit procedure. See *Appendix F* for a sample.

- Nominate a competent person to assume overall responsibility for ensuring the tasks involved in audit planning, conducting, reporting and the follow-up are implemented effectively.

- Prepare an audit plan that reflects the frequency and the scope of the audit. Also consider the importance of activities and results of previous audits.

- Select and train a team of auditors who could conduct audits. The competence of auditors plays an important role in the quality and output of audits, so it is worthwhile investing in auditor training and competence. A good auditor training programme could be spread over three to five days and include topics such as EMS issues, ISO 14001, auditing principles, audit communication and planning, conducting and reporting.

- Share the audit plan with auditors and auditees. Auditors must prepare themselves by gaining awareness about the auditee's processes and documentation in advance. Knowledge of applicable regulatory requirements and preparing checklists can significantly support an EMS audit.

- Audits are best conducted using a process-based approach. This shifts the focus from narrow, procedure-oriented hair splitting to the bigger EMS issues. In simple terms, an audit should gather evidence to determine the following:

  o Does the organisation meet ISO 14001's requirements?

  o Does the EMS enable the organisation to meet its environmental objectives and planned arrangements?

  o Does the EMS enable the organisation to prevent pollution and comply with legal and other requirements?

  o Is the EMS implemented effectively and continually improved?

- Auditors must be independent. You can ensure this by not nominating an auditor for an activity that they are partly or wholly responsible for.

- Auditors must be encouraged to adopt a friendly, professional and open approach to the audit process, and keep the auditee up to date with findings. Auditees prefer to have a sense of participation in the audit process and do not like surprises.

- A critical skill for an auditor is their ability to verify the effectiveness of corrective actions. To do this, they should consider:
  - o Have the root cause(s) been identified correctly?
  - o Will the actions taken ensure that the nonconformity will not reoccur?
  - o Have other similar situations, tasks, locations and equipment also been reviewed for removal of the same or similar nonconformity? A nonconformity is like a bad fish in a pond. Removing some and not others will surely spoil the remaining fish in due course.
  - o Have the issues of shortcomings in training, organisational structure, adequacy of resources and management commitment also been considered and addressed?

CHAPTER 11: MANAGEMENT REVIEW AND CONTINUAL IMPROVEMENT (CLAUSES 9.3 AND 10.3)

Summary of requirements

- It is the responsibility of top management to review the performance of its EMS at defined intervals.
- The review is intended to determine how well the system is performing and how it can be improved.
- The review must be based on inputs, such as audit results, legal compliance, communication(s) with interested parties, objectives, corrective and preventive actions, and recommendations for improvement.
- Top management should provide specific decisions and actions relating to continual improvement of the EMS.

How can these requirements be met?

'Management review' is the final platform where the EMS's performance and adequacy are reviewed. If an issue is not reviewed in the management review, there will be no other opportunity to properly address it.

Implementing a management review process requires the following:

- Define who constitutes top management. Normally, this comprises the most senior executives with final decision-making responsibilities.
- Define when management reviews take place. Avoid too frequent reviews (1 to 2 months) or too infrequent reviews (more than 12 months).
- Create a measuring and monitoring system covering the entire EMS's performance.
- The management representative collects and presents this data as input for the management review process. For a sample EMS management review input report, see *Appendix G*. An EMS input report is only as good as the data or facts that it contains. As such, an input report containing specific and accurate facts and results of the EMS's performance can greatly improve the effectiveness of the review process.
- It is good practice (though not a requirement) to share the EMS performance review report with top management before the actual review meeting. Top management reviews each element of the EMS's performance and determines its suitability, adequacy and effectiveness. Try to find opportunities for improvement, and necessary actions to realise those improvements.
- The EMS review process must result in decisions and actions relating to changes to correct any identified system deficiencies, or improve the EMS. A good management review should come up with clear

decisions on what, when and who shall take the necessary actions to make those corrections or improvements.

- ISO 14001 requires the organisation to maintain records of the EMS management review. These records could consist of the input report (see *Appendix G*) and the outcomes of the review in terms of decisions and actions taken by the management.

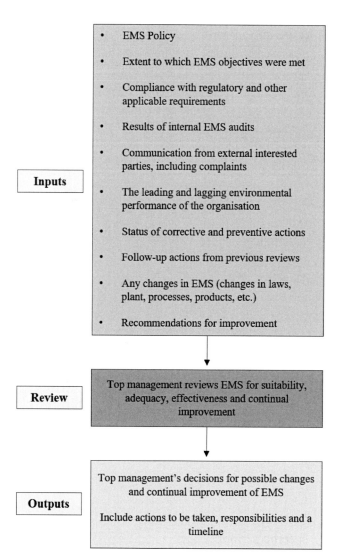

Figure 5: EMS management review process

Continual improvement

Although 'continual improvement' of the EMS is mandated by ISO 14001, it does not explicitly say how this can be achieved. Looking closely, however, several EMS elements suggest and provide input to the continual improvement process. Let's follow the trail of everything done so far in establishing the EMS and see how different elements create opportunities for continual improvement of the EMS.

- The EMS policy explicitly states top management's commitment towards continual improvement.
- Determining significant environmental aspects identifies those that require additional steps to be taken, hence improving the EMS.
- The demand for regulatory compliance requires an organisation to improve its environmental performance to keep itself well within the specified regulatory limits.
- Environmental objectives, targets and programmes provide a formal process of continual improvement.
- Operational controls help eliminate, reduce or prevent potential environmental impacts. Improving the effectiveness of operational controls or introducing additional controls is part of the continual improvement process.
- Measurement and monitoring processes provide specific performance data to identify deficiencies and realistic improvement targets.

- Corrective and preventive actions to avoid nonconformities reoccurring or developing allow for further improvement of the EMS.
- The management review process takes a holistic view of the EMS. It identifies deficiencies and considers possibilities for continual improvement.
- Continual improvement is the final step in the PDCA cycle. It reflects the true spirit of ISO 14001, which lies in not just establishing an EMS, but also adapting it to changing circumstances and continually making it more effective.

CHAPTER 12: NONCONFORMITY, CORRECTIVE AND PREVENTIVE ACTION (CLAUSE 10.2)

Summary of requirements

- Establish a system for responding to a real or potential nonconformity.
- Determine the cause(s) of nonconformities, take corrective or preventive actions (as applicable) and determine the effectiveness of actions taken. Mitigating environmental impact is an integral part of corrective actions.

How can these requirements be met?

An organisation may identify existing environmental nonconformities from any of the following sources:

- An internal or external environmental audit.
- Measuring, monitoring or testing for compliance or any other requirement.
- After an environmental incident.
- After a complaint from a regulatory body or other interested party.

An organisation may identify potential environmental nonconformities – opportunities for preventive actions – from any of the following sources:

- Data collection and analysis.
- New processes and products.

- Actions in response to leading EMS indicators (such as initiatives for environmental system improvement or upkeep, or regulatory change).

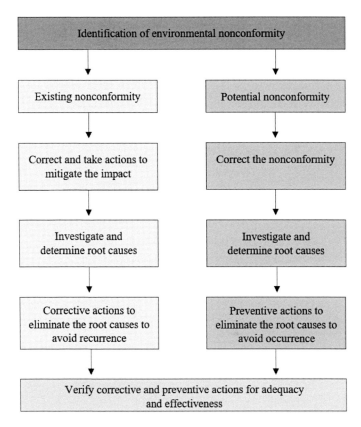

Figure 6: Corrective and preventive action flow

When identifying an existing or a potential nonconformity, take the following steps:

- Correct the nonconformity and mitigate environmental impacts – if any – through, for instance; containment, clean-up activities and compensation.

- Investigate and identify the root causes of the nonconformity; differentiate between immediate, direct and root causes. Consider, as an example, a fuel spill caused by the turning over of a fuel tank lorry. While the turning over of the fuel tank lorry is the immediate cause, rash driving could be the direct cause. A lack of training, inadequate knowledge of speed limits and absence of route risk assessment could be the root causes. Often there is more than one root cause that contributes to an incident, and all should be explored.

- Take actions to eliminate the root and direct cause(s) to prevent recurrence of the nonconformity.

- Verify actions taken for adequacy and effectiveness; problems may reoccur if the corrective actions are inadequate or ineffective. Verification of the effectiveness of corrective actions should be done by someone competent in root cause analysis tools, such as:
  o Cause and effect diagrams;
  o Why-why analysis;
  o Failure mode and effect analysis; and
  o Statistical tools.

- Maintain records of corrective/preventive actions.

Appendix E describes a suggested format for nonconformity reports.

CHAPTER 13: GREEN INITIATIVES

The term 'green initiative' could be explained as the schemes, projects or programmes undertaken to better the environment in order to eliminate or reduce the environmental footprint.

The entire ecosystem is facing a multitude of challenges caused by industrialisation, urbanisation and other activities that put pressure on the environment. Environmental degradation caused by pollution of air, water, land and atmosphere is considered a key reason for changes in climate, food shortages, extinction of flora and fauna, as well as many types of diseases. Industrialisation and commercial activities have been the biggest contributors to environmental degradation. The impacts are loud enough to convey a very strong message: environmental management is no longer a matter of choice. Organisations ought to consider it at each stage and for each business activity.

Green initiatives include efforts to eliminate, reverse or reduce the adverse environmental impacts of an organisation's products, processes, activities and services.

Figure 7 describes a model to develop an end-to-end green business strategy:

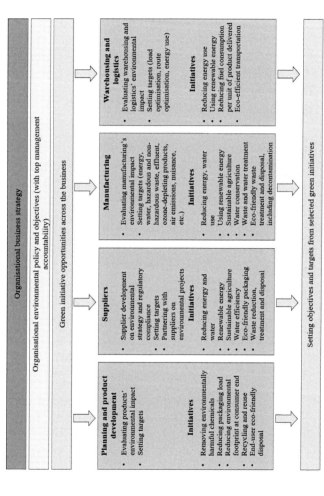

Figure 7: Green business strategy model

The following are examples of green initiatives.

IT industry:

- Energy-efficient computers, printers and other IT equipment with auto-sleep (low energy) mode when not in use.
- Developing products that run on solar energy.
- Recycling plastic, cartridges, toners, etc.

Packaging:

- Purchasing paper and board material only from suppliers that purchase their raw material from sustainable forests.
- Using eco-friendly packaging with 100% recycling and biodegradable options.
- Removing unnecessary packaging through improved design (e.g. air-shell packaging).

Offices and administration:

- Using renewable energy (solar panels).
- Adopting eco-friendly designs to use sunlight instead of artificial light.
- Using teleconferencing or videoconferencing to avoid road and air travel.
- Optimising off-hours energy use (only security, IT network and fire-alarm systems may need power backup, instead of the whole office).
- Switching to motion sensor lighting.

- Carpooling.
- Eliminating the use of disposable plastic bottles for drinking water.
- Encouraging the use of bicycles instead of company cars.

Warehousing and transportation:

- Designing warehouses to maximise the use of daylight.
- Using renewable energy (solar/wind).
- Reducing fuel consumed per unit of product delivered to customers by optimising routes or modes.

General:

- Eliminating lead from petrol.
- Investing in renewable energy projects (solar, wind, geo-thermal, etc.).
- Developing bio-fuels from agricultural waste or non-edible plantation (e.g. Jatropha).
- Establishing heat recovery projects (flare gases recovery and power generation projects, etc.).
- Engaging communities on environmental issues (awareness programmes, eco-friendly disposal techniques, reuse and recycling, drip irrigation techniques, reforestation programmes, discontinuing use of disposable plastic water bottles, etc.).

Consumer goods and foods:

- Developing products with fewer environmental impacts (detergents that consume less water while washing clothes, recyclable and/or biodegradable packaging, etc.).
- Reducing the use of natural resources (water, fossil fuels, etc.).
- Using 'pigging techniques' for waste and water reduction in changeovers.
- Eliminating marine pollutants from detergents, soaps and shampoos, etc., and replacing them with alternatives that are friendly to aquatic life.
- Replacing non-biodegradable plastic with biodegradable or other recyclable alternatives.
- Eco-friendly waste disposal (site decontamination during acquisitions and closures, used packaging collection and recycling system, etc.).

Miscellaneous industries:

- Using lead-free paints.
- Recovering and recycling chrome in the leather industry.
- Using ozone-friendly refrigerants.
- Replacing older vehicles with fuel-efficient ones.
- Using biomass in boilers, furnaces and for other heating requirements.

- Using cleaner fuels with lower emission impacts (e.g. replacing furnace oil boiler fuel with LPG or natural gas to reduce sulphur oxide emissions).
- Using waste/biomass for power generation.
- Carbon trading.

In short, green initiatives should become the core of an organisation's EMS. They help reduce the carbon footprint, prevent pollution and create a more sustainable environment. They enable an organisation to go beyond the minimum requirement of regulatory compliance and make a larger contribution to improve the environment and society.

APPENDIX A: SAMPLE PROCEDURE FOR IDENTIFYING ENVIRONMENTAL ASPECTS

Purpose

To define a process for identifying environmental aspects and impacts and determine those that have or can have significant environmental impacts. The procedure applies to all processes, products and activities carried out by Eco-Friendly Inc (EFI).

Responsibilities

Manager <   > is responsible for the following activities:

- Identifying environmental aspects, especially those with significant impacts.
- Maintaining the significant aspect register.

Procedure

1. Make an inventory of all processes, products and activities (routine, non-routine and emergency) performed by EFI.
2. Identify inputs and outputs of each process, product and activity using a black-box approach (as shown in *Figure 2*).
3. Identify environmental aspects associated with inputs and outputs. Consider the following sources while identifying aspects/impacts:

3.1. Observations, examination and brainstorming.

3.2. Process flow diagrams.

3.3. Material safety data sheets.

3.4. Data from previous incidents, accidents, spills, emissions, etc.

3.5. Views of interested parties (records of external communication).

3.6. Applicable legal and regulatory requirements.

3.7. Results of testing and analysis.

3.8. Input/output (mass balance).

3.9. Utility bills.

3.10. Waste hauling records.

4. Define the environmental impact of each aspect.

5. Rank each impact on a scale. While determining the scale of an impact, give due consideration to controls already in place, severity, frequency, legal requirements and reputational impact (see *Table 3*).

Table 4: Ranking Scale of Impacts

Significance criteria	Rating scheme
Legal liabilities: is the aspect regulated?	Low (1): not regulated High (4): regulated
Reputational concerns	Low (1) Medium (2) High (3) Very high (4)
Difficulty or cost of changing the impact	Low (1) Medium (2) High (3) Very high (4)
Frequency: what is the frequency of occurrence of this environmental aspect?	Low (1): could be once in 10 years Medium (2): less than once a year High (3): every few months Very high (4): occurs monthly or more frequently

Significance criteria	Rating scheme
Severity: what is the overall severity (scale, harm to environments, duration)?	Low (1) Medium (2) High (3) Catastrophic (4)
Degree of controls already in place	High degree of control (1) Medium degree of control (2) Moderate degree of control (3) Slight degree of control (4)

6. Use the format shown in *Table 4* for recording aspects and impacts.
7. Calculate the average of all six impact ratings. A score of two or above is considered a significant environmental impact.
8. Document aspects and impacts in a register of significant environmental aspects (see *Table 4*).
9. Review and keep updated the register of significant environmental aspects as an ongoing activity (at least once a year). Also review when new aspects are identified, an environmental incident takes place, laws are changed, or processes and activities are added or modified.

Records

The following records must be kept:

- A register of significant environmental aspects.
- Competence records for the environmental assessment team.

Table 5: Register of Significant Environmental Aspects

Activity/product/service				
Aspect				
Impact				
(a)	Degree of controls already in place			
(b)	Legal liability			
(c)	Reputational concerns			
(d)	Cost/difficulty of changing the impact			
(e)	Severity			
(f)	Frequency			
Overall score = Average (a:f) Significant if score > 2				

APPENDIX B: SAMPLE PROCEDURE FOR INTERNAL AND EXTERNAL COMMUNICATION

Purpose

To define a process for internal and external communication on environmental aspects and the EMS of Eco-Friendly Inc (EFI).

Responsibilities

Manager <   > is responsible for internal EMS communication.

Manager <   > is responsible for external EMS communication.

Procedure

1. Manager <   > ensures that a two-way communication process relating to environmental issues is established between personnel at all levels in the organisation. This includes visitors and contractors working on site. These processes shall ensure that the following EMS-related facts, reports, instructions and awareness procedures (as applicable) are communicated to all personnel:

   1.1. Environmental issues.
   1.2. Significant environmental aspects.
   1.3. Eco-friendly initiatives and environmental objectives.

1.4. Report environmental incidents by email.

1.5. Pollution prevention, waste management, recycling and reuse.

1.6. Emergency procedures.

1.7. Applicable operational controls, procedures and instructions.

1.8. EMS reports and results of EMS performance communicated to concerned managers and top management.

1.9. Applicable legal and other requirements are known to those responsible for measuring, monitoring or ensuring compliance.

2. There shall be at least two awareness sessions per year on each of the above topics, while the regulatory requirements and changes in procedures will be communicated immediately after receipt of a change.

3. Manager <   > ensures that a two-way communication process relating to external communication is established. The external communication process shall ensure receipt, response and recording of all communication relating to environmental issues. This may involve:

3.1. Communication with NGOs, environmental groups and communities;

3.2. Communication with government, regulatory and environmental protection agencies;

3.3. Communication relating to permits, licences and fines;

3.4. Communication with press and other media; and

3.5. Reporting of mandatory environmental data to environmental agencies and corporate bodies.

Records

Records that must be kept include records of receipts and responses to all external communications.

APPENDIX C: SAMPLE PROCEDURE FOR CONTROL OF DOCUMENTED INFORMATION

Document no. EFI-01

Revision: 0

Revision date: dd/mm/yyyy

Purpose

This procedure defines the mechanism for developing, approving, reviewing, updating, distributing and maintaining controlled documents used in EFI's EMS. It also includes mechanisms for identifying, preserving, maintaining, retaining and disposing records.

Responsibility

Manager <   > has the overall responsibility to ensure implementation of this procedure.

Procedure

Management of controlled documents

1. All EMS documents shall bear a unique identification number following the pattern EFI-01, EFI-02, EFI-03, and so on.
2. Manager <   > shall allocate a unique number to each document.
3. Manager <   > shall ensure that a master list (EFI-02) of all EMS-controlled documents is maintained,

identifying each document and its current revision status. The same will be done for documents of external origin.

4. Manager <   > shall ensure that the documents are distributed such that they are available and accessible to everyone who needs to use them.

5. EMS documents at EFI shall be approved by the following personnel:

Table 6: Document Approval Responsibilities

Document title	Approving authority
EMS manual	Company head < >
EMS policy	Company head < >
EMS objective and targets	Company head < >
EMS programmes	Manager < >
EMS procedures	Manager < >
Standard operating procedures (or work instructions)	Manager < >
EMS forms for recording results and/or data	Manager < >

6. Documents are assigned a revision number and date to indicate their latest revision status.

7. Those issued with controlled documents must ensure that they remain available and legible.

8. Manager < > shall ensure that obsolete documents are promptly removed from all points of issue. Obsolete documents will be segregated to prevent unintended use. Manager < > shall decide if an obsolete document is to be destroyed or retained for historical purposes. If retained, the document shall be stamped as 'obsolete document' to prevent unintended use.

9. Changes in documentation may be initiated by anyone who identifies the need for or proposes a change.

10. Changes to documents must be reviewed and approved by the same authority that had reviewed and approved the original document.

11. Changes in documents shall be made easily noticeable by writing the changed text in ***bold and italic*** font.

Management of records

12. The retention period of each record and the responsibility for record maintenance is defined at EFI-03. The legal and organisational requirements

are considered while establishing retention periods of records.

13. Each record holder is responsible for ensuring that the records are readily available and legible. They are kept safe and prevented from loss or damage.

14. Records must be traceable by including information such as events, dates, persons and approvals.

15. Records that have completed their retention period are disposed of. Record holders may revise retention periods, where necessary, and changes are made to EFI-03 accordingly.

16. The following controls are applicable for electronically maintained records:

    16.1. Manager <   > shall ensure that access and authorisation is controlled through passwords and/or access codes.

    16.2. Data loss is prevented by ensuring regular backups and data corruption is prevented by an online antivirus scanning program.

    16.3. Manager < > shall ensure that data is backed up and maintained at a remote location on data storage devices.

Records

Table 7: Example Record Maintenance

Record no.	Record name	Maintained by	Retention period
EFI-02	List of documents	Manager < >	2 years
EFI-03	List of records	Manager < >	2 years

Approved by: ....

APPENDIX D: DOCUMENTED INFORMATION REQUIRED BY ISO 14001

Clause	Documents required by the Standard
4.3	Scope of the EMS
5.2	Environmental policy
6.1.1	Risks and opportunities that the organisation needs to address
6.1.1	Processes required by the risk assessment, environmental aspects, compliance obligations
6.1.2	Environmental aspects and associated impacts
6.1.2	Criteria used to determine significant environmental aspects
6.1.2	Significant environmental aspects
6.1.3	Compliance obligations
6.2.1	Environmental objectives
7.2	Evidence of competence

Clause	Documents required by the Standard
7.4	Evidence of internal and external communications
7.5.1	Documented information determined to be necessary for the EMS
8.1	Documented information necessary to have confidence that EMS processes have been carried out as planned
8.2	Emergency preparedness and response
9.1.1	Evidence of monitoring, measurement, analysis and evaluation of the EMS
9.1.2	Evidence of evaluation of compliance
9.2.2	Evidence of the internal audit programme and results
9.3	Evidence of the results of management reviews
10.2	Evidence of the nature of nonconformities and subsequent actions taken
10.2	Evidence of the results of corrective actions

APPENDIX E: SAMPLE FORMAT FOR A NONCONFORMITY REPORT (NCR)

NCR no.:	NCR date:
Reference no. of document against which this nonconformity is raised: company procedure / ISO 14001 / legal requirement / complaint from interested party / other:	
Describe nonconformity: Signed: Date:	

Action taken to correct the nonconformity (include actions taken to mitigate environmental aspects, if any):

Signed:

Date:

Root cause(s):

Actions taken to eliminate root cause(s) to avoid recurrence:

Signed:

Date:

Review of corrective actions for effectiveness:

Signed:

Date:

APPENDIX F: SAMPLE PROCEDURE FOR INTERNAL EMS AUDIT

Purpose

To define the process and responsibilities for planning, conducting, reporting and following up internal EMS audits at Eco-Friendly Inc (EFI).

Responsibilities

Manager <   > has the overall responsibility for:

- Planning internal EMS audits;
- Ensuring auditor competency and independence;
- Designating auditors for specific audits;
- Ensuring regulatory compliance is also audited;
- Monitoring how audits are conducted; and
- Following up and reporting the results of audits to senior management.

Procedure

Planning

1. All departments of EFI shall be audited against all elements of the Standard at least once a year. Additional audits may be scheduled considering the environmental significance of activities and the results of previous audits.

2. Manager < > shall prepare an audit plan for each audit and communicate it to the auditors and auditees at least three weeks before the audit.
3. The audit plan defines:
   3.1. The audit scope and criteria;
   3.2. The audit team members and functions to be audited;
   3.3. Audit dates, times and any other requirements; and
   3.4. Responsibilities for writing the audit report and taking follow-up actions.

Responsibilities/qualifications

4. Manager < > shall nominate a lead auditor and at least one additional auditor as the audit team for each audit. The auditors would be considered suitable to conduct an internal EMS audit if they meet the following conditions:
   4.1. They meet auditor competence requirements defined in the training procedure of EFI.
   4.2. They are independent of the activity and/or department being audited.
   4.3. They have participated in at least two audits as an observer auditor.

Internal EMS audit process

5. EMS audits will be conducted primarily through interviews, observing physical conditions and checking documents and records.

6. The audit team will conduct short opening and closing meetings with the departmental head of the area that is being audited to explain the scope and methodology of the audit.

7. Auditors will document nonconformities and hand them over to the management of the area under audit. A copy of the nonconformity report is given to Manager <   > to record and follow up.

8. Responsibility for determining root cause(s) and taking corrective actions will reside with the functional area managers where findings occurred. Management representatives should be requested to coordinate where required.

9. The lead auditor (unless otherwise specified by Manager <   >) should follow up to ensure that the corrective actions are completed by the agreed-upon dates. The EMS audit is closed once establishing that the corrective actions are fully implemented and effective.

Audit report

10. When the audit is complete, the lead auditor will complete the audit report and make it available to

Manager <   > and the relevant functional area manager.

Audit as input to the management review

11. Manager <   > shall present the results of the audit and the status of the corrective actions to the top management of EFI as input for the management review process. Any nonconformities that have not been fully implemented will continue to be monitored, and a status report will be provided until the corrective action is fully implemented.

Records

Records that must be kept include:

- An audit plan;
- Audit nonconformity reports; and
- The audit report.

APPENDIX G: SAMPLE INPUT REPORT FOR EMS MANAGEMENT REVIEW

Input report for the management review to be conducted on 12 July 2018 by the top management of Eco-Friendly Inc (EFI).

Extent to which EMS objectives and targets were met

EFI had established two EMS objectives at the beginning of 2018. These were:

- **The water consumption in the factory shall be reduced from 15,000 gallons per month to 12,000 gallons per month by December 2018**
  Status: this objective has been fully achieved and our current monthly water consumption has been reduced to 11,000 gallons per month.
- **Reduction of carbon emissions by 2% by December 2019**
  Status: six targets were identified for achievement of this objective: three for 2018 and three for 2019. The three targets for 2018 (carbon emission study, procurement of carbon reduction equipment and optimising the combustion process of the two boilers) have been achieved.

Compliance with regulatory and other applicable requirements

EFI complies with all applicable regulatory requirements relating to emissions, effluents, solid waste and permit renewals. This was reconfirmed in the six-monthly compliance audit conducted in June 2018. Test results showing EFI performance (emissions, effluents, hazardous waste and permits) against applicable regulatory requirements are attached with this report.

Results of internal EMS audits

Two EMS audits were carried out in 2018: a complete system audit in February and an EMS audit of the production department in May.

Complete system audit – February 2018

It was a four man-day audit carried out by trained internal auditors. The audit confirmed that EFI was in complete compliance with all regulatory requirements. The system was well implemented, and the objectives and targets were being met. The audit also identified one nonconformity. Details are given below:

- **Nonconformity:** scrubber used for reducing emission of sulphur oxide in Chimney No. 2 was inoperative for the past five weeks.
- **Root cause:** ineffective maintenance programme and lack of awareness among the scrubber maintenance personnel.

- **Corrective actions taken:**
  - o Scrubber repaired and reinstalled.
  - o All other scrubbers' performance checked.
  - o The maintenance system was reviewed; it was ensured that all pollution prevention equipment would be identified and a maintenance plan made to ensure ongoing preventive maintenance (*Procedure EFI-07, Rev. 2* was revised to include the new requirements).
  - o Training need identification was revisited and the training requirements for pollution prevention equipment were identified for all maintenance personnel (the new training requirements were included in the 2018–2019 training calendar).

Production department audit – May 2018

It was a two man-day audit carried out by trained internal auditors for the production department only. The audit confirmed that EMS operational controls were well in place. The audit also identified one nonconformity. Details are provided below:

- **Nonconformity:** no operational controls deployed for potential spillage or leak from oil drums kept in store.
- **Root causes:**
  - o Inadequate assessment of significant impacts in case of emergency situations.
  - o Inadequate measuring and monitoring.

- o Lack of awareness of operational controls.
- **Corrective actions taken:**
  - o Tiled surface prepared for areas where drums are kept to cater for drainage of potential spillage and leakage.
  - o Inspection of incoming drums introduced to separate any damaged or leaking drums.
  - o Purchase process modified (*EFI-05, Rev. 4*).
  - o Training need identification was revisited and the training requirements for EMS operational controls were identified for all concerned persons (new training requirements were included in the 2018–2019 training calendar).
  - o Other lube oil storage locations were inspected to ensure that similar controls are also implemented in those locations.
  - o Spill kits placed in each lube oil storage location to be used for containment in case of spills (personnel also trained in the use of spill kits).

Communication from external interested parties, including complaints

Since the last management review meeting there was one complaint and one query received from interested parties. The communication received and actions taken were as follows:

- **Communication:** complaint received from local road-traffic control department on 10 June 2018. The complaint related to spillage and leaks from EFI's vehicles while travelling between EFI and the customers' premises.

Action: EFI raised a nonconformity report. Lack of vehicle maintenance and inadequate inspection of vehicles before departure from the factory were identified as the root causes. Corrective actions were taken to introduce regular vehicle inspections as well as pre-departure vehicle checks. A response was sent to the traffic control department explaining all the corrective actions taken by EFI.

- **Communication:** 12 April 2018, 'The Green Initiatives', an NGO working towards environmental protection, sent a query asking if the effluent discharged by EFI met the biological oxygen demand (BOD) limits of 30 mg/litre.

  Action: as it is the policy of EFI to disclose its environmental performance (if specifically asked for by an interested party), a response was sent stating that the BOD value of EFI effluent is 25 mg/litre, which is well within the stated limits.

Environmental performance

The environmental performance as judged from lagging environmental performance indicators.

Table 8: Lagging EMS Performance Indicators

Lagging parameters (based on 50,000 tonnes of production from July 2017 to June 2018)	Performance
Amount of hazardous waste generated	2,000 kg
Quantity of toxic chemicals released	Nil
Number of notices of violation	Nil
Amount of fuel used	5,000 gallons
Amount of water used	11,000 gallons per month
Ozone-depleting substances in use (relate to old versions of air-conditioning units)	6
Hazardous air pollutants (HAPs) released to air	20 pounds

Spills of oil or hazardous substances (notified to regulatory bodies and corrective actions taken)	1
Energy usage/unit of product	0.23 kilowatt hour

Table 9: Leading EMS Performance Parameters

Leading parameter	Performance
Number of purchase reviews completed for environmental aspects	3
Number of voluntary initiatives participated in	2
Number of community outreach activities	2 half-day sessions
Number of internal self-assessments completed	2
Number of environmental management reviews completed	2

Regulatory issues identified proactively and resolved	1	
Number of EMS training courses completed	9	

Status of corrective and preventive actions

- Corrective actions initiated: 3
  Current status: all closed out. Results verified for effectiveness.
- Preventive actions initiated: 1
  Current status: action pending because of delay in budget approval.

Follow-up actions from previous reviews

Previous review decisions included two items:

- Revision of contractor management procedure to include extending EMS controls to those contractors working for EFI outside company premises. Completed through *EFI-09, Rev. 3*.
- Building a wall to contain generator noise from disturbing the neighbourhood residents. This has been accomplished and the neighbouring community is fully satisfied with the new arrangement.

Any changes in EMS or circumstances that could relate to environmental aspects

Two changes have been introduced that have a potential for new or changed environmental aspects in the organisation. These are:

- Constructing a new warehouse for storage of paints and epoxies; and
- Introducing 50 new printers that could generate hazardous solid waste (toners and cartridges) in the weeks to come.

EMS policy

The EMS policy of EFI was last revised in June 2017. The policy is enclosed for review and possible changes by top management.

Recommendations for improvement

Three recommendations are being made for consideration by top management for continual improvement of the EMS. These are:

- Making a new EMS objective for reducing the carbon footprint of the organisation;
- Providing training on the corrective and preventive action process to all managers; and
- Purchasing BOD/COD testing equipment so that the testing of outgoing effluent can be carried out more frequently to exercise ongoing control over these parameters.

Signed:

Management representative (EMS)

BIBLIOGRAPHY

ISO 14001:2015 Environmental management systems – Requirements with guidance for use: *www.iso.org/standard/60857.html*.

ISO 19011:2018 Guidelines for auditing management systems: *www.iso.org/standard/70017.html*.

FURTHER READING

IT Governance Publishing (ITGP) is the world's leading publisher for governance and compliance. Our industry-leading pocket guides, books, training resources and toolkits are written by real-world practitioners and thought leaders. They are used globally by audiences of all levels, from students to C-suite executives.

Our high-quality publications cover all IT governance, risk and compliance frameworks and are available in a range of formats. This ensures our customers can access the information they need in the way they need it.

Other resources you may find useful include:

- *ISO 14001 2015 EMS Documentation Toolkit*
  www.itgovernancepublishing.co.uk/product/iso-14001-2015-ems-documentation-toolkit
- *A Guide to Effective Internal Management System Audits*
  www.itgovernancepublishing.co.uk/product/a-guide-to-effective-internal-management-system-audits
- *Business Management Controls - A Guide*
  www.itgovernancepublishing.co.uk/product/business-management-controls

For more information on ITGP and branded publishing services, and to view our full list of publications, visit *www.itgovernancepublishing.co.uk*.

To receive regular updates from ITGP, including information on new publications in your area(s) of interest, sign up for our newsletter at *www.itgovernancepublishing.co.uk/topic/newsletter*.

Branded publishing

Through our branded publishing service, you can customise ITGP publications with your company's branding. Find out more at *www.itgovernancepublishing.co.uk/topic/branded-publishing-services*.

Related services

ITGP is part of GRC International Group, which offers a comprehensive range of complementary products and services to help organisations meet their objectives.

For a full range of resources on ISO 14001 visit *www.itgovernance.co.uk/shop/category/iso-14001*.

Training services

The IT Governance training programme is built on our extensive practical experience designing and implementing management systems based on ISO standards, best practice and regulations.

Our courses help attendees develop practical skills and comply with contractual and regulatory requirements. They also support career development via recognised qualifications.

Learn more about our training courses and view the full course catalogue at *www.itgovernance.co.uk/training*.

Professional services and consultancy

We are a leading global consultancy of IT governance, risk management and compliance solutions. We advise businesses around the world on their most critical issues and present cost-saving and risk-reducing solutions based on international best practice and frameworks.

We offer a wide range of delivery methods to suit all budgets, timescales and preferred project approaches.
Find out how our consultancy services can help your organisation at *www.itgovernance.co.uk/consulting*.

Industry news

Want to stay up to date with the latest developments and resources in the IT governance and compliance market? Subscribe to our Daily Sentinel newsletter and we will send you mobile-friendly emails with fresh news and features about your preferred areas of interest, as well as unmissable offers and free resources to help you successfully start your projects.
www.itgovernance.co.uk/daily-sentinel.

ND - #0263 - 120824 - C0 - 198/129/7 - PB - 9781787780323 - Gloss Lamination